MANAGING CLIMATE RISKS IN RESILIENT CITIES

MANAGING CLIMATE RISKS IN RESILIENT CITIES

Lawrence Susskind

THE UNIVERSITY OF UTAH PRESS

Salt Lake City

Publication of this keepsake edition is made possible in part by
The Wallace Stegner Center for Land, Resources and the Environment
S.J. Quinney College of Law
and by
The Special Collections Department
J. Willard Marriott Library

This lecture was originally delivered on March 30, 2016, at the 21ˢᵗ annual symposium of
the Wallace Stegner Center for Land, Resources and the Environment.

The Defiance House Man colophon is a registered trademark
of The University of Utah Press. It is based upon a four-foot-tall Ancient Puebloan
pictograph (late PIII) near Glen Canyon, Utah.

21 20 19 18 17   1 2 3 4 5

Cover and frontispiece photo "From the High Line" courtesy of Iker Alonso.

FOREWORD

The Wallace Stegner Lecture serves as a public forum for addressing the critical environmental issues that confront society. Conceived in 2009 on the centennial of Wallace Stegner's birth, the lecture honors the Pulitzer prize–winning author, educator, and conservationist by bringing a prominent scholar, public official, advocate, or spokesperson to the University of Utah with the aim of informing and promoting public dialogue over the relationship between humankind and the natural world. The lecture is delivered in connection with the Wallace Stegner Center's annual symposium and published by the University of Utah Press to ensure broad distribution. Just as Wallace Stegner envisioned a more just and sustainable world, the lecture acknowledges Stegner's enduring conservation legacy by giving voice to "the geography of hope" that he evoked so eloquently throughout his distinguished career.

The 2016 Wallace Stegner Lecture was delivered by Professor Lawrence Susskind from the Massachusetts Institute of Technology on the subject of "Managing Climate Risks in Resilient Cities." Given that climate change impacts are already evident, Professor Susskind exhorts local communities to begin taking action to adapt to climate risks and urges them to employ a new mode of public engagement, one that puts the responsibility on citizens to devise ways to manage these risks that not only meet their own self-interest but the interests of others as well. The message is both timely and provocative.

Robert B. Keiter, Director
WALLACE STEGNER CENTER FOR LAND,
RESOURCES AND THE ENVIRONMENT

MANAGING CLIMATE RISKS IN RESILIENT CITIES

Preparing for and managing climate change–related risks will require whole communities to act, because individuals are limited in what they can do to protect themselves. This idea is at the heart of what I am going to be talking about: whole communities need to act in order to ensure that those communities become resilient in the face of climate-related risks. Think of flood emergencies. You can try to flood-proof your house and put your most precious things higher up in case water comes into the basement, but you can't ensure that the electricity in the whole community will come back on, water and sanitation systems will work, roads will be open, first responders will arrive when you need them, hospitals will operate, food will be delivered, and public health standards will be restored. So if you have a flood, there is only so much you can do as an individual.

There are steps that local residents and decision-makers can take together to manage possible climate risks such as increased storm intensity, sea level rise (which we are concerned about on the coasts), increasing numbers of very hot days, loss of winter snowpack, and periods of drought. In general, I think of climate risks in terms of too much or not enough water in the wrong place at the wrong time and extended periods of higher than usual temperatures. When I talk with communities on both coasts about climate change, basically what we are talking about is water and heat.

The chance that the climate is changing should be an important factor in every city and town's decisions about its long-term development planning, including infrastructure investment, the administration of zoning and building codes, and investments in open-land preservation. Just like communities prepare for possible water shortages (whatever their cause) or earthquakes, coastal and riverine communities should be working to reduce their vulnerability and enhance their resilience in the face of climate risks, and look to do both in ways that maximize a range of cobenefits (i.e., achieving other things that are important to the community at the same time). If we enhance emergency preparedness, for example, it will provide benefits regardless of the type of emergency that occurs. So, whether you think climate change has immediate or even long-term effects, enhancing emergency preparedness has cobenefits. We might act because we are concerned about flooding, but the same emergency warnings, practice evacuations, beefing up of medical assistance, and training for first-responders will pay off if there's a hazardous spill of some kind. In general, no matter how great the uncertainty, I would argue that risk management investments pay off.

There are basically three ideas I want to talk about today regarding the management of climate risks:

1. Communities should focus now on adaptation, not just mitigation.

   I'll explain these two terms for anyone not already familiar with that distinction. If you focus on adaptation first, I believe, you'll build a clear case for mitigation over time. Adaption means look at the risks now; mitigation means deal with the underlying causes. So, if you reduce CO_2 emissions, presumably you will reduce the long-term effects of climate change. But all the things you have to do to reduce CO_2 emissions (which take a long time) are different from what you can do right now to reduce community vulnerability and enhance community resilience to the kinds of risks I have listed related to climate change. Communities should focus on adaptation, not just mitigation.

2. Communities will have to take everybody's views into account, not just the advice of technical experts, if they are going to get anything accomplished. There are too many places where people are treating climate-related risks as technical problems and assuming that if people with technical knowledge would tell us what to do, then everyone would do it. We can see very clearly that is not going to happen,

which means you better find a way to take account of everyone's views if you expect to take any action on adaptation.

3. First, though, you must bring everyone up to speed. Don't just talk about risk management, do something. Take small steps, send the right signals, and encourage cooperative efforts.

In the city where I live, Cambridge, Massachusetts, we just spent another million dollars on a climate vulnerability study. Honestly, there's going to be too much water or not enough water in the wrong place at the wrong time and a lot more hot days in a row. I do not need to spend a million dollars to figure that out. How much more water, which place, how many hot days, when? I'm not convinced spending money on studying that right now is going to allow us to get a really good fix because the uncertainties involved are too great. I'd rather take the million dollars and spend it on enhanced emergency preparedness. I'd rather do something about reducing flood risk. So, my concern is doing something that actually sends signals to people demonstrating that there are ways of managing those risks.

FOCUS ON ADAPTATION

Let me discuss the adaptation point first. Most of the conversation about climate change—all over the world—has been focused on reducing greenhouse gas emissions, especially the CO_2 that is a byproduct of energy production. This is called mitigation. Climate action plans in many cities call for increased energy efficiency, reductions in the use of fossil fuels, and a shift to renewables— climate action plans aimed at reducing CO_2 emissions. There is a great deal more that each community can do to reduce CO_2 emissions, but there are a great many forces pushing in the opposite direction. Oil prices are dropping, not increasing. Many current industries are unable or unwilling to invest in new technologies;

they are operating right now at the edge of profitability. Anything that involves new capital investment in the short term, even if they agree completely in the value of reducing CO_2—they just can't do it. Some people think they have decades before the effects of climate change will hit, and they certainly don't want to make investments now if they can be put off. They say, "Well, it's not really clear and it's long term. We have very short-term, immediate demands and problems. We can't do those things you're saying to achieve mitigation." And, some people believe that human-induced climate change is a hoax and they certainly don't want to see any activity moving public resources in a direction that suggests that climate change is real, because they are convinced it's not. When pressed at the most recent global climate change summit in Paris, a great many countries were unwilling to commit to the scale of CO_2 reduction that most scientists believe is required to hold off global warming. So there are all kinds of pressures pushing against the investment that it would take to reduce long-term climate risks.

At the same time, small island nations around the world already see the effects of sea level rise. Drought, expansion of desert areas, increasing storm intensity, more forest fires, extensive flooding, migration of flora and fauna, heat island effects in cities, and changes in disease vectors are already evident—measured, not forecasted. In response to the horrible impacts of Katrina and Superstorm Sandy, a great many coastal and riverine communities have initiated studies of their climate change vulnerabilities and have begun to sort out ways of enhancing their resilience. This is called adaptation (in contrast to mitigation, which is aimed at reducing CO_2 emissions).

There is no reason that a community can't focus on both mitigation and adaptation at the same time, but given all the other pressing needs in most cities that doesn't seem likely. The point I want to make today is that emphasizing adaptation now may be the best way of convincing residents to take action on mitigation in the years ahead. When communities see how much it is going to cost them, year after year, to reduce their vulnerability to the impacts of sudden climate change and enhance their resilience in the face of floods, drought, sea level rise, and extended heat spells (especially insofar as managing public health impacts are concerned), they will eventually ask how they can get at the root of the problem. The answer, of course, is reducing CO_2 emissions in the long run, and that's mitigation. But people aren't there yet and, given that they aren't there, a focus on adaptation makes real the everyday costs of coping with climate change–related risks.

I believe that public investment in short-term climate risk management efforts (to reduce vulnerability and enhance resilience) will help to build long-term public support for mitigation, which appears to be lacking at present. Also, the most effective risk management efforts in the short-term are likely to yield a wide range of cobenefits that help communities achieve other important objectives simultaneously. And while we've tried to make that point about mitigation, it's a much harder point to make.

No effort towards adaptation is going to happen unless there's community support for the kind of expenditure and regulatory changes involved. I don't care who the next elected leader is; they're not going to get anything done unless there is public support for taking these kinds of actions.

Seek Widespread Public Engagement

Local risk management efforts that can help communities avoid or respond to the impacts of sudden climate change include:

Emergency preparedness measures.
1. Public education about actions individual property owners can take (i.e., storm-proofing).
2. "Hardening" of basic utilities and infrastructure. If you live along the coast of Massachusetts like I do, every time there's a storm the power goes out. Why? Well, the power lines are right along the coast, so when a storm comes in it knocks them over. You say, "Why don't you bury those lines?" Oh,

it's too expensive. What does it cost to keep replacing them year after year? Well, that's an annual cost. If you spent the money to bury the system, you wouldn't have to keep doing that year after year.

We have one case of a city north of Boston where one storm came in and destroyed the water treatment facility. Why? Where's the water treatment facility? Down by the harbor. Why? Because that's where they own some land and didn't have to buy an additional piece of land. So the storm came in and destroyed the facility. They still have fifteen years left on the bond and have to keep paying that, and then they have to get another bond to build a new facility. Where are they going to build it? Same location. Why? Because they would have to spend more money to get another piece of land and move the pipes over there. So we bond again, build again, and next year's storm destroys the system. Now we have to pay the bonds on two nonusable facilities. So the mayor says maybe we shouldn't build it a third time in the same place, but a group of businesses in the harbor file a lawsuit saying that it would take too long to get another site up and they would lose business because they need the water treatment facility running. They go to court and the mayor is enjoined from speaking about an alternative location; the justice department requires her to take out another bond and rebuild the facility as quickly as possible. I know hardening basic utilities and infrastructure has its costs, but it is a way communities can adapt to risk.

3. Encourage retreat or abandonment (with compensation) from properties in the most vulnerable areas. We know this would save lives in the long term, reduce costs in the long term, and put fewer first responders at risk in emergencies. But the notion of a policy that tells private landowners to retreat (i.e., to move from where they are because that area is most vulnerable) is a very, very hard sell—even with the offer of compensation, people don't want to move. They do

expect first responders to show up at public expense when they are stuck, once again, in that spot in the middle of a storm. But they argue that retreating from or abandoning private property should not be something they are required to do.

4. Adoption of new public health requirements, building codes, and land use regulations.
5. Offering financial incentives to individual property owners to take resiliency-enhancement measures.

Which of these efforts to take at various points in time is a question that entails public policy choices that are the responsibility of elected and appointed officials. But, because each involves the distribution of gains and losses to different groups (at different times), they need public support. The only way to generate that support is through widespread public engagement.

Public engagement can take many forms. Traditionally, public information campaigns, public meetings (i.e., neighborhood forums), surveys, or public advisory committees have been used to ensure notice and opportunities to respond to what government has in mind. Civil society organizations, of course, often run their own parallel efforts (some along the same lines). Sometimes companies or individuals use advertising or social media campaigns to coalesce public concern about the risk management choices I have listed.

These can be divided into two general categories: information out and public reactions back. The first seeks to "educate" the public along the lines that experts and officials think best. The second aims to document public opinion, but does nothing to ensure that citizens are working with reliable and relevant information, or that they get to hear what others think and want.

There is a third form of public engagement that we advocate in our new book *Managing Climate Risks in Coastal Communities: Strategies for Engagement, Readiness, and Adaptation* (see more at localclimatechange.mit.edu). This third form of public engagement

involves face-to-face problem-solving forums. Assuming you want
to get people's attention, give them information they can trust and
use to weigh their options. Bring them together with their neighbors
who have different opinions to see if they can reach agreement on
the trade-offs that need to be made. Face-to-face means it's not an
auditorium, it's not a survey—it requires people face-to-face with
people who have different views from their own to decide what
they would recommend to government.

When we do this we build it around what are called role-play
simulations, or games, that convey a great deal of factual material
quickly and painlessly and bring people together in small two-hour
facilitated consensus-building workshops to hear how their neighbors would try to meet their own needs as well as the needs of
others. We have helped to organize such sessions in several parts
of the country and we've measured what happens when you put
together a game where people come and sit together at tables of
seven or eight. We try to mix them by the way they identify their
interests and they are given a role to play in a hypothetical community a lot like their own with basic data about this community.
We've taken some time to put that information together by consult-
ing with different groups and organizations so that people in the
group are likely to say it looks appropriate, and then we ask people
to play a role not their own. So, if you're an environmentalist you're
asked to play a local industrial role; if you're a government person,
you are asked to play the community activist role, and you're given
a short, one-page set of confidential instructions that came from
interviewing people in that real role in that place.

In this game you are given an assignment and a set of choices
to make within an adaptation strategy, with some indication of the
likely costs and benefits. The group you are part of is given an hour
to see if people can reach agreement, as group members try to stay
consistent with the instructions they've been given (and which
they've had about twenty minutes to read). Afterwards, we talk
about it and we get very different results from different tables playing
the same game at the same time. We talk together to see what the

source of the disagreement was, what type of arguments carried weight, and to what extent we can summarize the types of arguments that people would recommend. We put that information together and share it with public officials trying to make that information public and even stream the conversations.

It's not that the decisions made at that table are telling the city what to do; the most important thing is that people are getting a sense of how their views differ from those of others in their community. They are also getting an opportunity to think about joint problem-solving responsibility—they explore how what they want and need can mesh with what other people want and need. The face-to-face forums demonstrate that there is a possibility for the community to reach agreement on how to manage climate risks.

When we survey people before they sit down and after the forums, we see a rather stark difference in their sense of the urgency of climate-related risks. We don't ask, "Do you believe in climate change?" We say: "If flood risks are increased over the next few years, which of these strategies do you think we should pursue to deal with that?" or "If there are going to be extended heat waves, and we know people die in those heat waves if they don't have mobility, what are the things that the hypothetical community can do? What do you suggest? Can you reach agreement?" The only results accepted from the groups are agreements, not a summary of how they disagreed—which is what most public engagement normally gets you, thereby letting those who are in elected power do whatever they want with impunity since there was no agreement amongst all the people giving feedback. Our research demonstrates that this form of public engagement, especially if it generates agreement among diverse sets of informed stakeholders, provides elected and appointed officials with the push they need to adopt climate risk management measures.

This third form of public engagement puts the responsibility on citizens to think about ways of managing climate risks that not only meet their own interests but the interests of others as well. And, it gives them reliable local forecasts of the range of possible

short-term, mid-term, and long-term changes in average precipita-
tion, sea level rise, and temperature in their area using a downscaled
local model of climate forecasting. It also provides a sense of the
probable costs and benefits of the various actions that communities
can take.

Start with Small Symbolic Steps

I want to emphasize the concept of cobenefits. If storms have already
threatened the integrity of sewage treatment facilities, power plants,
water supplies, the electric grid, and transportation facilities, efforts
to harden or defend these public services make sense regardless of
the motivation for making improvements. We have communities
on the East Coast that are hard hit every year by winter storms.
Their power systems and water systems are repeatedly destroyed.
They rebuild them in the least expensive way over and over again.
When someone suggests that perhaps they ought to spend more
and build a more appropriately fortified (or buried) version of the
same facilities, local budget pressures argue against taking such
action. So, there are cities that bond the same facility repeatedly,
and are now paying off multiple bonds for a facility that they have
to build yet again.

Climate risk management might provide the impetus to rebuild
in a smarter way. This is a good thing.

In some communities we talk about flood risks (whatever their
cause) and try to think about ways of adapting to one-hundred-year
floods that seem to be happening every few years. A one-hundred-
year flood, by the way, is not a flood that happens once every
hundred years. Rather, it is a flood that—based on past experi-
ence—has a 1 percent chance of occurring every year. Climate
change is modifying these odds.

I live on a small lake—we own the dam, we are responsible for
the safety of the dam, and we are now responsible for managing
the dam in the face of a risk of a five-hundred-year flood. I say,
"What's the risk of a five-hundred-year flood? Show me the data.

How many five-hundred-year floods have we had? None? Why are we planning for that?" Because it's more likely than it used to be. "But it's expensive to rebuild the dam." Yes, but you have to build to a five-hundred-year flood risk. It's very difficult to convince my neighbors to do this, but otherwise we can't get insurance. Without it, the risk to everything else downstream goes up. Communities are now talking about smarter flood risk management (regardless of what is causing the flooding). But they are only ready to take such action when there is a shared commitment and a widespread understanding of the nature of the risks, costs, and benefits of alternative risk management actions. That's why public engagement that aims to enhance readiness to make decisions together is so important.

Communities would do well to keep the public informed even after agreement has been reached on a first round of climate risk management efforts. I think every taxpayer wants to hear what the return is on their collective investments.

Joint monitoring and continued adjustment ought to be the order of the day. Imagine that every newspaper, radio station, and TV station reported changing climate risks to each community every few months. How have vulnerability levels gone down this month or this year? How have resilience levels gone up? What's made a difference? We have to make people conscious of the fact that they are making risk management decisions already, and most of the time they are asking the government to do nothing in regard to risk management.

In our most recent research at the MIT Science Impact Collaborative, we have found that people respond differently to climate risk when floods, storms, and heat are translated into possible public health impacts. When the sewage system is overwhelmed by a flood and local drinking water is contaminated, a health impact assessment (HIA) can show residents what kinds of illnesses and diseases they face. You only need to read about Flint, Michigan, to understand that when you translate climate risk into public health risk people have a different level of concern. When a string of very hot days occurs, we can demonstrate how cases of renal failure, especially among the elderly, are likely to increase. When floods leave ponds to still water, we can show that an increase in diseases associated with waterborne insects might occur. We haven't seen vulnerabilities of climate change expressed in public health terms, and that is what we are now trying to do: help communities translate climate risks into public health risks. When you talk about the gains and losses associated with different risk management strategies, you can see it in terms of public health advances.

When communities understand the public health risks of heat, drought, and floods, the cobenefits of managing climate-related risks will look much more attractive.

About the Author

Lawrence Susskind has served in a variety of roles at the Massachusetts Institute of Technology (MIT) over the past four decades, including as head of the Department of Urban Studies and Planning (DUSP), and founder and current head of DUSP's Environmental Policy and Planning Group. He teaches full time at MIT, advises a large group of doctoral students, and supervises a number of research teams at the MIT Science Impact Collaborative (SIC), including the MIT-UTM Malaysia Sustainable Cities Program. The MIT SIC is a long-term research enterprise focusing on innovative approaches to public engagement in resource management and sustainable development. Currently MIT SIC is working on hydropower in Southern Chile, energy development in Mexico, transboundary water disputes in Africa, and sustainable city development in Malaysia. He founded and is currently chief knowledge officer of the Consensus Building Institute, a Cambridge-based nonprofit that provides mediation and negotiation training services in many parts of the world. Professor Susskind was one of the founders, twenty-five years ago, of the interuniversity Program on Negotiation (PON) at Harvard Law School. He is currently vice-chair for instruction at PON and offers executive training including a Master Class in Negotiation and Advanced Mediation Training for Lawyers. He codirects the MIT-Harvard Public Disputes Program that focuses on land rights of indigenous peoples around the world, negotiation of transboundary water disputes, resolution of disputes over sacred land, and the resolution of science-intensive policy disputes. He has produced more than fifty role-play simulations and numerous teaching videos. Professor Susskind is the author or coauthor of twenty books, including, most recently, *Managing Climate Risks in Coastal Communities* (2015), *Environmental Diplomacy* (Second Edition, 2014), *Good for You, Great for Me* (2014), *Water Diplomacy* (2012), and *Breaking Robert's Rules* (2006). His books have been republished in eight languages.